巨型恐龙
supersize Dinosaurs

[英] 露丝·欧文/著

刘颖/译

汉英对照
恐龙科普

江苏凤凰美术出版社

全家阅读
小贴士

★ 每天空出大约10分钟来阅读。

★ 找个安静的地方坐下，集中注意力。关掉电视、音乐和手机。

★ 鼓励孩子们自己拿书和翻页。

★ 开始阅读前，先一起看看书里的图画，说说你们看到了什么。

★ 如果遇到不认识的单词，先问问孩子们首字母如何发音，再带着他们读完整句话。

★ 很多时候，通过首字母发音并听完整句话，孩子们就能猜出单词的意思。书里的图画也能起到提示的作用。

最重要的是，感受一起阅读的乐趣吧！

扫码听本书英文

Tips for Reading Together

• Set aside about 10 minutes each day for reading.

• Find a quiet place to sit with no distractions. Turn off the TV, music and screens.

• Encourage the child to hold the book and turn the pages.

• Before reading begins, look at the pictures together and talk about what you see.

• If the child gets stuck on a word, ask them what sound the first letter makes. Then, you read to the end of the sentence.

• Often by knowing the first sound and hearing the rest of the sentence, the child will be able to figure out the unknown word. Looking at the pictures can help, too.

Above all enjoy the time together and make reading fun!

Contents 目录

十分高大
Very Tall

许多恐龙的体形都很大，但有些恐龙尤其高大。

这种十分高大的恐龙是长颈巨龙。

它能吃到大树顶端的叶子。

Lots of dinosaurs were big but some dinosaurs were very, very big.

This very tall dinosaur is a Giraffatitan.

It could eat leaves from the top of a tall tree.

长颈巨龙英文名的字面意思是"巨大的长颈鹿"。

The name Giraffatitan means giant giraffe.

长颈巨龙 Giraffatitan
(ji-RAF-ah-tie-tun)

比公交车长
Longer Than a Bus

梁龙的体长相当于3辆公交车!

它是一种食草动物。

它能吃很多很多植物。

Diplodocus was as long as three buses!

It was a **herbivore**.

It ate lots and lots of plants.

它每隔4周就会换新牙。
Its teeth wore out after four weeks and new teeth took their place.

梁龙 Diplodocus
(di-PLOD-u-kuss)

巨大的化石
Giant Fossils

我们之所以认识这些巨大的恐龙，是因为科学家发掘出了它们的化石。

We know about these huge dinosaurs because **scientists** dig up their **fossils**.

恐龙尾巴化石
fossil dinosaur tail

2005年，科学家在南美洲发现了145块巨大的化石。

In 2005, scientists found 145 huge fossils in South America.

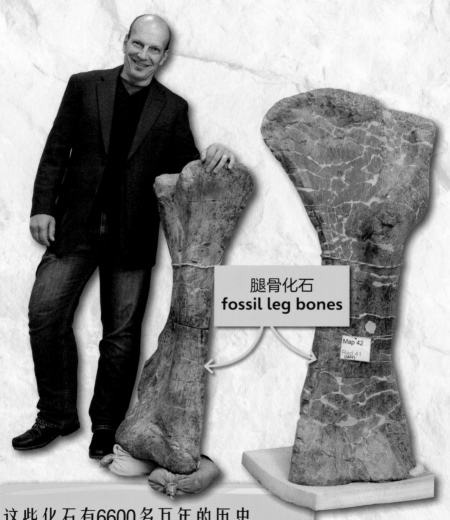

腿骨化石
fossil leg bones

这些化石有6600多万年的历史。
The fossils were more than 66 million years old.

一种新恐龙
A New Dinosaur

科学家像拼图一样将骨头化石拼在一起。
他们发现了一种新的巨型食草性恐龙。

Scientists fitted the fossil bones together like a jigsaw puzzle.
They had found a new kind of huge, plant-eating dinosaur.

这些白色的骨头就是科学家发现的145块化石。
The white bones are the 145 fossils found by the scientists.

科学家称这种恐龙为无畏龙。
The scientists called the dinosaur Dreadnoughtus.

无畏龙
Dreadnoughtus
(DRED-nawt-us)

认识无畏龙
Meet Dreadnoughtus

无畏龙从头到尾长26米。

Dreadnoughtus was 26 metres long
from its head to the end of its tail.

26 米
26 metres

它的脖子比公交车还长。

Its neck was longer than a bus.

它和12头大象一样重！

It weighed the same as 12 elephants!

最大的恐龙
The Biggest Dinosaur

2014年，科学家发掘出迄今为止最大的恐龙化石。

它的体长接近4辆公交车。

In 2014, scientists dug up the fossils of the biggest dinosaur that has ever been found.

It was nearly as long as four buses.

它是在巴塔哥尼亚发现的，因此被称为巴塔哥巨龙。

It was found in Patagonia and it is called Patagotitan.

巴塔哥巨龙
Patagotitan
(patter-go-TIE-tun)

植物和粪便
Plants and Poo

巨型恐龙几乎都以植物为食。
科学家之所以知道这一点，是因为他们
发现了这些恐龙的粪便化石。

The biggest dinosaurs were plant-eaters.
Scientists know this because they have found fossils
of their poo.

富塔隆柯龙
Futalognkosaurus
(foo-TAL-ung-ko-SAW-rus)

粪便中有一些树叶、细枝、草和蕨类植物。

Inside the poo are bits of leaves, twigs, grass and ferns.

粪便化石
poo fossil

认识棘龙
Meet Spinosaurus

棘龙的体长超过了1辆公交车。

它背上的棘和人一样高。

棘龙牙齿的形状很像鳄鱼的牙齿。

Spinosaurus was longer than a bus.

The **spines** on its back were as tall as
a man.

It had teeth like a crocodile.

棘龙的牙齿
Spinosaurus tooth

18

棘龙英文名的字面意思是"有棘的蜥蜴"。
Spinosaurus means "spine lizard".

棘 spines

棘龙 Spinosaurus
(SPY-no-SAW-rus)

最长的肉食恐龙
The Longest Meat-Eater

棘龙是最长的食肉性恐龙。

同鳄鱼一样，棘龙可在陆地上和水里生存。

Spinosaurus was the longest of the
meat-eating dinosaurs.

It lived in water and on land, like a crocodile.

它背上的棘被皮肤覆盖着。
The spines on its back were covered in skin.

棘龙 Spinosaurus

词汇表 Glossary

化石　fossil

存留在岩石中几百万年前的动物和植物的遗体。

The rocky remains of an animal or plant that lived millions of years ago.

食草动物　herbivore

只吃植物的动物。

An animal that only eats plants.

科学家　scientist

研究自然和世界的人。

A person who studies nature and the world.

棘　spine

动物身上坚硬而锋利的刺突。

A hard, sharp point on an animal's body.

恐龙小测验 Dinosaur Quiz

① 梁龙有多长？

How long was Diplodocus?

② 每隔4周，梁龙的牙齿会怎么样？

What happened to Diplodocus teeth after four weeks?

③ 科学家是如何发现巨型恐龙的？

How do scientists find out about the supersize dinosaurs?

④ 棘龙哪里像鳄鱼？

How was a Spinosaurus like a crocodile?

⑤ 你想遇见梁龙还是棘龙？

Would you rather meet a Diplodocus or a Spinosaurus?